# Learn To Weld

## Essentials Of Metal Fabrication And MIG Welding For Beginners With Detailed Practical Techniques And Illustrations

## Gideon Ankunding

## Table of Contents

# Introduction

Welding is a widely used process in manufacturing, construction, and various other industries. It involves joining two or more pieces of metal together by melting the materials at the joint and allowing them to cool, thereby creating a strong and permanent bond. Welding is a fundamental skill that has revolutionized the way structures and products are built.

The process of welding involves several key elements, including a heat source, a filler material (in some cases), and protective measures. The most common heat sources used in welding are electric arcs, flames, lasers, or even friction. The choice of heat source depends on the specific application and the materials being welded.

During the welding process, intense heat is generated, causing the materials to melt and fuse together. This fusion creates a metallurgical bond between the pieces, resulting in a joint that is as strong or sometimes even stronger than the original materials. The use of a filler material, such as a welding rod or wire, may be necessary to provide additional strength and fill any gaps between the pieces being joined.

Welding also requires careful consideration of safety precautions. The high temperatures and intense light emitted during welding can pose risks to the welder's eyes and skin. Therefore, protective gear, such as welding helmets, gloves, and clothing, is essential to ensure personal safety.

There are various welding techniques available, each suited for different applications and materials. Some common welding methods include:

Shielded Metal Arc Welding (SMAW): Also known as stick welding, SMAW uses a consumable electrode coated in flux. The electrode melts and forms a protective gas shield around the weld, preventing contamination.

Gas Metal Arc Welding (GMAW): Also known as MIG/MAG welding, GMAW uses a continuously fed wire electrode and a shielding gas. The wire melts and forms the weld, while the gas shield protects the molten metal from the atmosphere.

Tungsten Inert Gas Welding (TIG): TIG welding uses a non-consumable tungsten electrode to produce the arc and a separate filler material if needed. An inert gas, such as argon, is used to shield the weld from the surrounding air.

Flux-Cored Arc Welding (FCAW): FCAW is similar to GMAW, but it uses a tubular electrode filled with flux. The flux produces a shielding gas when heated, eliminating the need for an external shielding gas.

These are just a few examples of welding techniques, and there are many others, each with its own advantages and applications.

Welding plays a crucial role in numerous industries, including construction, automotive, aerospace, shipbuilding, and manufacturing. It enables the fabrication of complex structures and components, ensuring strength, durability, and reliability.

Mastering welding techniques requires training, practice, and an understanding of different materials and their properties. Skilled welders are in high demand and can pursue various career paths, ranging from construction and fabrication to inspection and quality control.

# CHAPTER ONE

## What is Welding?

Welding is a manufacturing process that involves joining two or more pieces of metal together by melting the materials at the joint and allowing them to cool, thereby creating a strong and permanent bond. It is the process of fusing metals through the application of heat and, in some cases, the use of a filler material.

The primary objective of welding is to create a strong and continuous joint between the materials being joined.

When the metal pieces are heated to their melting point, they become molten and form a pool of liquid metal. As the molten metal cools and solidifies, it creates a metallurgical bond between the pieces, resulting in a cohesive joint.

The heat required for welding can be generated using various sources, such as electric arcs, flames, lasers, or even friction. The choice of heat source depends on factors such as the type of metal, the thickness of the materials, and the specific welding technique being used.

In some welding processes, a filler material is used to facilitate the joining of the metal pieces. The filler material, which is often a metal alloy with a lower melting point than the base metals, is added to the joint to provide additional strength and fill any gaps or voids between the pieces being joined.

Welding is a versatile process that can be used to join a wide range of metals and alloys, including steel, aluminum, stainless steel, copper, and titanium, among others. It is employed in various industries, including construction, automotive, aerospace, shipbuilding, and manufacturing, to create structures, components, and products.

To ensure the safety and quality of welding operations, proper training, adherence to safety precautions, and knowledge of different welding techniques and materials are essential. Skilled welders possess expertise in selecting the appropriate welding method, adjusting parameters such as heat input and welding speed, and ensuring the integrity of the welds.

# History of Welding

The history of welding dates back thousands of years, with evidence of early welding techniques found in ancient civilizations.

Bronze Age and Ancient Times: The earliest evidence of welding comes from the Bronze Age, around 3000 BCE, where copper and bronze were joined using a technique known as forge welding. This involved heating the metals and hammering them together to create a bond. Similar techniques were used in ancient times by civilizations such as the Egyptians, Greeks, and Romans.

Middle Ages and Renaissance: During the Middle Ages, blacksmiths used forge welding to join iron and steel. The process involved heating the metals in a forge and then hammering them together. In the 16th century, the discovery of the blowpipe allowed the use of a directed flame for heating, enabling more precise control over the welding process.

19th Century: The 19th century witnessed several significant advancements in welding. In 1800, Sir Humphry Davy, an English chemist, invented the electric arc, which provided a new heat source for welding. However, practical applications of electric arc welding were not realized until the late 1800s.

Late 19th to Early 20th Century: In the late 1800s, Nikolay Benardos, a Russian inventor, and Stanislaw Olszewski, a

Polish inventor, independently developed the concept of carbon arc welding, which utilized a carbon electrode and an electric arc. This technique was further improved by C.L. Coffin in the United States, leading to the commercialization of carbon arc welding.

Early 20th Century: The early 20th century saw the development of new welding processes. In 1903, French engineer Auguste De Meritens invented the first practical electric arc welding process using a coated metal electrode. This paved the way for shielded metal arc welding (SMAW), also known as stick welding, which became widely used.

World War I and World War II: Welding gained significant importance during both World Wars. The demand for joining metal components quickly and efficiently led to advancements in welding technology. Welding was extensively used in shipbuilding, aircraft manufacturing, and the production of military equipment.

Post-World War II: The post-war period saw further developments in welding technology. In the 1940s, gas metal arc welding (GMAW), commonly known as MIG/MAG welding, was introduced. It utilized a continuously fed wire electrode and a shielding gas to protect the weld from atmospheric contamination.

Late 20th Century to Present: The latter half of the 20th century saw the emergence of various advanced welding techniques. Tungsten inert gas welding (TIG) and plasma arc welding (PAW) were developed, providing precise control over the welding process. Laser beam welding and electron beam welding also gained popularity, offering high-energy heat sources for welding.

Today, welding continues to evolve with advancements in automation, robotics, and materials science. The development of new welding techniques, equipment, and safety measures has greatly expanded its applications across industries, making it an essential process for modern manufacturing and construction.

# CHAPTER TWO

## Importance of Welding in Various Industries

Welding plays a crucial role in various industries, contributing to the manufacturing and construction of numerous products and structures. Some industries where welding is of utmost importance are as follows:

Construction: Welding is widely used in the construction industry for the fabrication and assembly of steel structures, such as buildings, bridges, and infrastructure projects. It allows for the efficient joining of structural components, ensuring strength and durability in the final construction.

Automotive: Welding is integral to the automotive industry for the manufacturing of vehicles. It is used to join metal components, such as body panels, frames, and exhaust systems. Welding techniques like spot welding and laser welding are employed to create strong and precise bonds, contributing to the structural integrity and safety of automobiles.

Aerospace: Welding plays a critical role in the aerospace industry, where precision and reliability are paramount. It is used to fabricate aircraft components, such as fuselage structures, engine parts, and fuel systems. Advanced welding techniques, like electron beam welding and laser welding, are employed to meet the stringent requirements of the aerospace sector.

Ship-building: Welding is essential in shipbuilding for the construction of vessels, offshore platforms, and marine structures. It enables the joining of metal plates and sections to form hulls, decks, and other ship components. The welding process ensures watertight and structurally sound connections, enabling the safe operation of ships and maritime structures.

Manufacturing: Welding is a key process in manufacturing industries, including machinery, appliances, and consumer goods production. It is used to fabricate metal components, assemble parts, and create intricate structures. Welding allows for efficient production, customization, and the assembly of complex systems.

Oil and Gas: Welding is vital in the oil and gas industry for the construction and maintenance of pipelines, storage tanks, and refineries. Welded joints in these critical infrastructures must withstand high pressures, corrosive environments, and extreme temperatures. Welding techniques and materials are carefully chosen to ensure the integrity and longevity of these structures.

Energy and Power Generation: Welding is utilized in the energy sector for the construction of power plants, including nuclear, fossil fuel, and renewable energy facilities. Welding is essential for fabricating boilers, pressure vessels, heat exchangers, and piping systems. The ability to create reliable and leak-free welded connections is crucial for the safe and efficient operation of power generation facilities.

Defense and Military: Welding is extensively used in defense and military applications, where the fabrication of armored vehicles, weapons, and military equipment requires strong and durable joints. Welding ensures the structural integrity, performance, and reliability of defense systems.

# Welding Processes

Arc welding is a welding process that utilizes an electric arc as the heat source to melt and join metals together. Within arc welding, there are different variations, including Shielded Metal Arc Welding (SMAW) and Gas Metal Arc Welding (GMAW), also known as MIG (Metal Inert Gas) welding.

Shielded Metal Arc Welding (SMAW): Shielded Metal Arc Welding, commonly known as stick welding, is one of the oldest and most versatile arc welding processes. It involves the use of a consumable electrode coated with a flux that melts and forms a protective gas shield around the weld area. The flux coating also provides additional protection against atmospheric contamination, improving the quality of the weld. SMAW is widely used for construction, maintenance, and repair applications due to its portability and ability to weld various metals and thicknesses.

Gas Metal Arc Welding (GMAW/MIG):

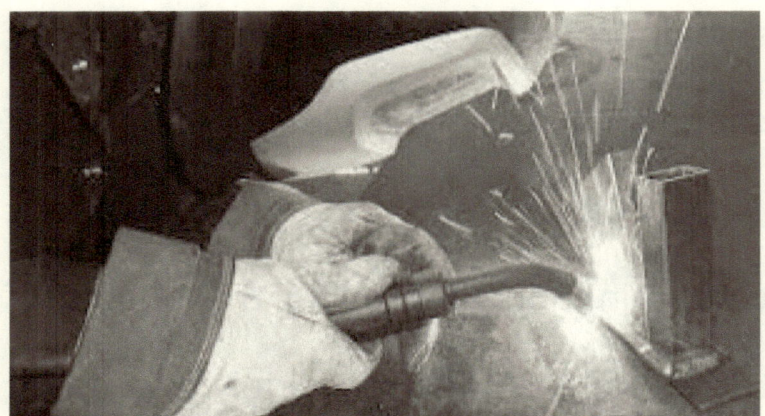

Gas Metal Arc Welding, often referred to as MIG (Metal Inert Gas) welding or GMAW (Gas Metal Arc Welding), is a popular arc welding process that utilizes a continuous solid wire electrode and a shielding gas. The electrode wire and shielding gas are fed through a welding gun, and an electric arc is established between the wire and the workpiece, melting both the wire and the base metal. The shielding gas, typically a mixture of argon and carbon dioxide, protects the molten metal from atmospheric contamination. GMAW is known for its speed and versatility and is commonly used in automotive, manufacturing, and fabrication industries.

These two arc welding processes are widely employed in various industries due to their effectiveness, versatility, and relatively simple operation. Both SMAW and GMAW offer advantages such as the ability to weld a wide range of materials, good weld quality, and the option to weld in different positions. However, they also have specific considerations and techniques that need to be mastered for optimal results. Other variations of arc welding include Flux-Cored Arc Welding (FCAW) and Gas Tungsten Arc Welding (GTAW/TIG), which have their unique applications and characteristics.

Gas Tungsten Arc Welding (GTAW/TIG):

Gas Tungsten Arc Welding, commonly known as TIG (Tungsten Inert Gas) welding or GTAW (Gas Tungsten Arc Welding), is a precise and versatile welding process that uses a non-consumable tungsten electrode to create an electric arc. The electrode does not melt during the welding process and serves as a heat source, while a separate filler metal can be added manually if needed. A shielding gas, typically argon or a mixture of argon and helium, is used to protect the weld from atmospheric contamination. GTAW/TIG welding is known for its ability to produce high-quality, clean welds in various metals, including stainless steel, aluminum, and copper. It is commonly used in applications that require precise control over the welding process, such as aerospace, automotive, and artwork fabrication.

Flux-Cored Arc Welding (FCAW):

Flux-Cored Arc Welding, abbreviated as FCAW, is an arc welding process that uses a tubular electrode filled with flux instead of a solid wire electrode. The flux provides a shielding gas when heated, protecting the weld from atmospheric contamination. FCAW can be performed with or without an external shielding gas, depending on the specific type of flux-cored wire used. This welding process is popular in industries such as construction, shipbuilding, and structural fabrication, as it allows for high deposition rates and can be used on thick materials. FCAW is particularly useful in outdoor or windy conditions since the flux creates a self-shielding effect, reducing the sensitivity to environmental factors.

Resistance Welding:

Resistance Welding is a welding process that joins metals by applying pressure and passing an electric current through the materials to generate heat. The heat is generated by the resistance to current flow at the joint between the workpieces, causing localized melting and subsequent fusion. Resistance welding methods include Spot Welding, Seam Welding, and Projection Welding, among others. These techniques are widely used in automotive manufacturing, appliance production, and the fabrication of sheet metal components. Resistance welding offers advantages such as high production speeds, automation capability, and the ability to join dissimilar metals. It provides strong and reliable welds with minimal distortion.

Each of these welding processes offers unique advantages and applications, catering to specific welding requirements and materials. The selection of the appropriate welding process depends on factors such as the type of materials being joined, the desired weld quality, production requirements, and environmental considerations.

Oxyfuel Welding:

Oxyfuel Welding is a welding process that uses a mixture of fuel gases and oxygen to generate a flame for heating and melting the metals being joined. The most common fuel gases used in oxyfuel welding are acetylene, propane, and natural gas. The heat from the flame melts the base metal, and a filler rod may be added to create a weld joint. Oxyfuel welding is commonly used for welding thin materials, such as sheet metal, and in applications that require portability and simplicity. It is often utilized in maintenance and repair work, as well as in the plumbing and HVAC industries.

Laser Welding:

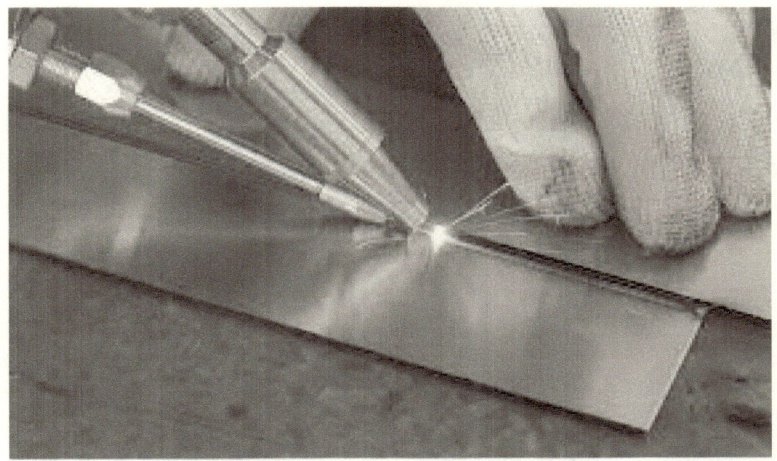

Laser Welding is a high-precision welding process that uses a focused laser beam to melt and join metals. The laser beam provides an intense heat source that rapidly melts the material, forming a weld joint. Laser welding offers several advantages, including precise control over heat input, minimal distortion, and the ability to weld highly reflective materials. It is commonly used in industries such as automotive, electronics, aerospace, and medical device manufacturing. Laser welding can be performed in a variety of configurations, including spot welding, seam welding, and remote welding, providing versatility and flexibility in applications.

Electron Beam Welding:

Electron Beam Welding (EBW) is a

welding process that uses a highly concentrated beam of high-velocity electrons to melt and join metals. Electrons are accelerated to high speeds and directed onto the workpiece, generating intense heat upon impact. The heat from the electron beam melts the metal, creating a fusion weld. Electron beam welding offers deep penetration and narrow weld profiles, making it suitable for joining thick materials with a high aspect ratio. It is commonly used in industries such as aerospace, automotive, and power generation, where precision and deep weld penetration are required. Electron beam welding is typically performed in a vacuum environment to prevent the scattering of electrons.

# CHAPTER THREE

## Other Welding Processes

Plasma Arc Welding (PAW):

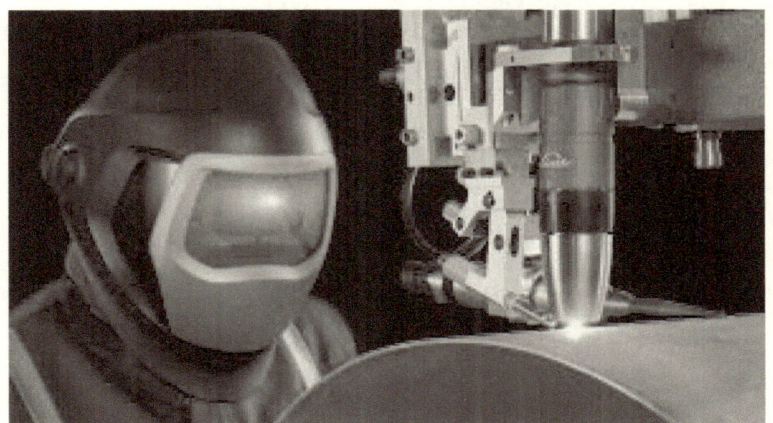

Plasma Arc Welding is a welding process
similar to TIG welding but uses a more
focused plasma arc. The plasma arc is
created by passing an electric current
through a gas passing through a
constricted nozzle. The concentrated
plasma arc provides higher heat

intensity, allowing for faster and deeper weld penetration. PAW is commonly used for precision welding in industries such as aerospace, automotive, and electronics.

Friction Stir Welding (FSW):

Friction Stir Welding is a solid-state welding process that joins materials by using frictional heat and mechanical pressure. A rotating tool with a specially designed pin is inserted between the materials to be joined. As the tool moves along the joint, it generates heat and stirs the material, creating a solid-state bond. FSW is particularly useful for joining materials that are difficult to weld with traditional fusion welding methods, such as aluminum and other non-ferrous alloys. It is widely used in the aerospace, automotive, and shipbuilding industries.

Ultrasonic Welding:

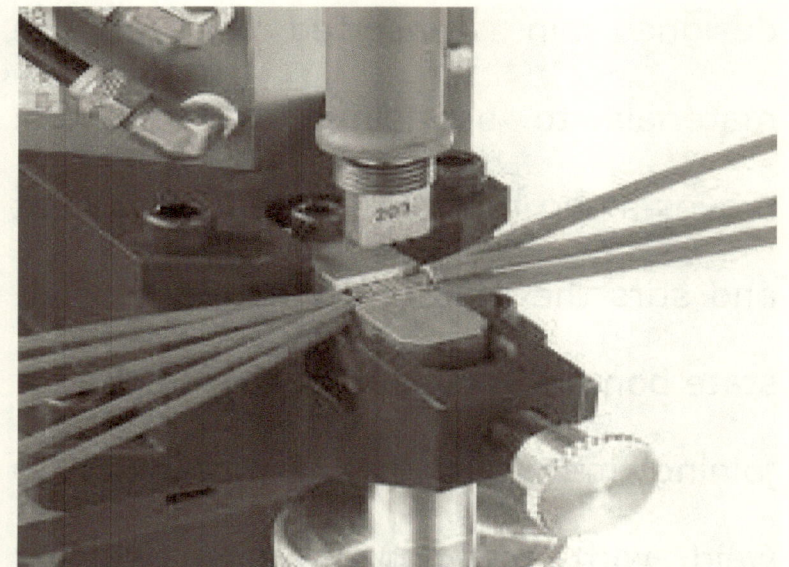

Ultrasonic Welding is a solid-state welding process that uses high-frequency ultrasonic vibrations to create frictional heat at the joint between two materials. The materials to be joined are held together under pressure while ultrasonic vibrations are applied, causing localized melting and bonding.

Ultrasonic welding is commonly used for joining plastics and is prevalent in industries such as electronics, automotive, and medical device manufacturing.

Electron Beam Welding (EBW):

As mentioned earlier, Electron Beam Welding uses a highly concentrated beam of high-velocity electrons to melt and join metals. It offers deep penetration and narrow weld profiles, making it suitable for thick materials and high-quality welds. EBW is commonly used in industries such as aerospace, automotive, and power generation.

Magnetic Pulse Welding (MPW):

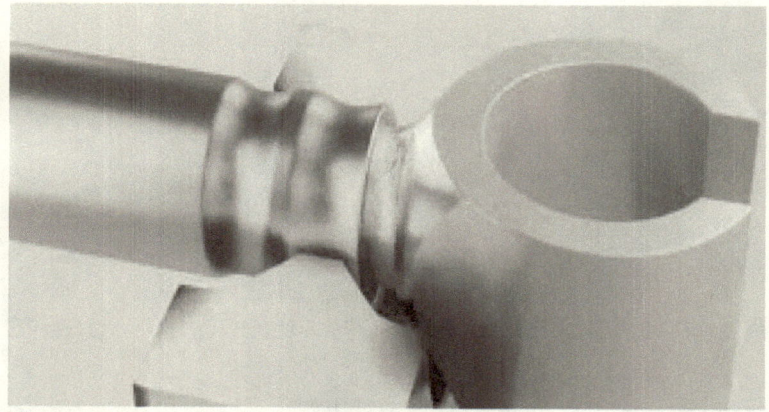

Magnetic Pulse Welding is a solid-state

welding process that uses magnetic forces to create high-velocity impact between two workpieces. A pulsed magnetic field accelerates one workpiece towards the other, generating a collision and a rapid plastic deformation that leads to a solid-state weld. MPW is commonly used for joining dissimilar metals and can produce high-quality welds without the need for filler material or heat-affected zones. It finds applications in industries such as automotive, aerospace, and electronics.

# Welding Equipment and Tools

Welding Machines and Welding Power Sources are key components of the equipment used in various welding processes.

Welding Machines:

Welding machines are devices that provide the necessary power and control for the welding process. They supply the electric current required to create an arc or heat source for melting and joining the metals. Different types of welding machines are designed for specific welding processes. Some common types of welding machines include:

Arc Welding Machines:

These machines are used for arc

welding processes such as Shielded Metal Arc Welding (SMAW), Gas Metal Arc Welding (GMAW/MIG), and Gas Tungsten Arc Welding (GTAW/TIG). They generate the appropriate current and voltage required for the specific welding process.

Resistance Welding Machines:

Resistance welding machines are used for resistance welding processes like Spot Welding and Seam Welding. They provide the necessary pressure, current, and timing control to create welds through the resistance heating of the workpieces.

Laser Welding Machines:

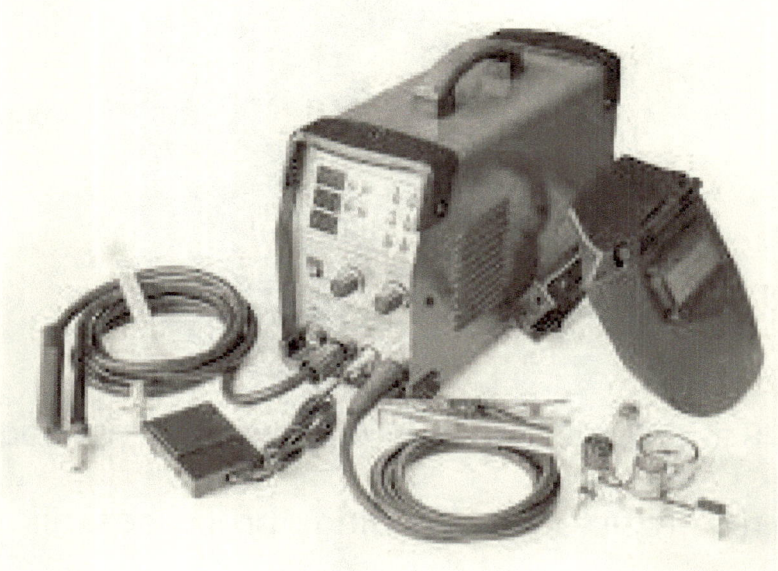

Laser welding machines utilize lasers to generate the high-intensity beam required for laser welding. These machines include laser sources, focusing optics, and control systems for precise beam control.

Electron Beam Welding Machines:

Electron beam welding machines are specialized equipment designed for electron beam welding processes. They generate and control the high-velocity electron beam required for the welding process.

Welding Power Sources:

Welding power sources are the devices that supply the electrical power to the welding machine. They convert input power from the electrical grid or other power sources into the appropriate output power required for welding. The type of power source used depends on the welding process and the specific requirements of the application. Some common types of welding power sources include:

Transformer-based Power Sources:

These power sources use transformers to step down the voltage from the electrical supply and provide the necessary current for welding. They are commonly used in traditional welding processes such as SMAW.

Rectifier-based Power Sources:

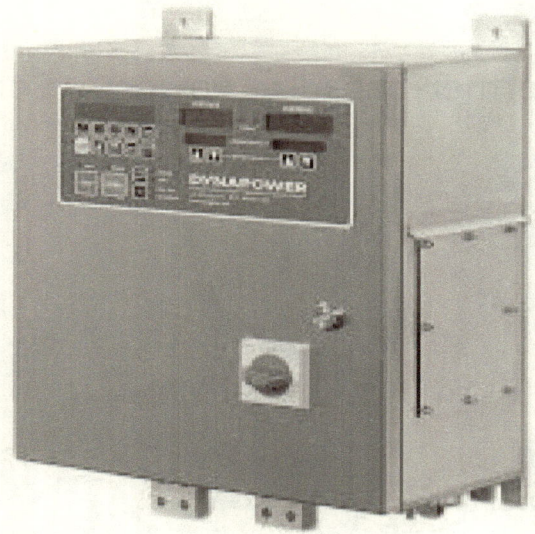

Rectifier-based power sources convert

alternating current (AC) from the electrical supply into direct current (DC) suitable for welding. They are used in processes such as GMAW/MIG and GTAW/TIG welding.

Inverter-based Power Sources: Inverter-based power sources use advanced electronics to convert AC power to high-frequency DC power. They offer benefits such as higher efficiency, portability, and improved control over the welding process. Inverter technology is commonly used in various welding processes, including SMAW, GMAW/MIG, and GTAW/TIG.

Capacitor Discharge Power Sources: Capacitor discharge power sources are used in certain resistance welding processes, such as Capacitor Discharge Spot Welding. They store electrical energy in capacitors and release it rapidly to create high-intensity welds.

These welding machines and power sources are essential for providing the required electrical current, voltage, and control to perform various welding processes. The selection of the appropriate equipment depends on factors such as the welding process, the type of materials being welded, the desired weld quality, and the specific application requirements.

# CHAPTER FOUR

## Electrodes and Filler Materials

In welding, electrodes and filler materials are used to facilitate the joining of metals and to enhance the quality and strength of the weld. Here's a breakdown of these components:

Electrodes: Electrodes are consumable materials that conduct electricity and deliver the welding current to the workpiece. The electrode can either be consumable, meaning it melts and becomes part of the weld, or non-consumable, where it remains intact during the welding process. The type of electrode used depends on the welding process and the materials being joined. Some common types of electrodes include:

Stick Electrodes: Stick electrodes, also known as welding rods, are used in Shielded Metal Arc Welding (SMAW). They consist of a metal core wire coated with a flux material. The flux provides a protective gas shield, removes impurities, and stabilizes the arc during welding.

Tungsten Electrodes: Tungsten electrodes are non-consumable electrodes used in Gas Tungsten Arc Welding (GTAW/TIG). They are made of pure tungsten or a tungsten alloy and are highly resistant to heat. Tungsten electrodes are used to generate the electric arc and do not melt during the welding process.

Filler Materials: Filler materials, also called welding consumables, are additional metal materials added to the weld joint to enhance its strength, integrity, and properties. The filler material must be compatible with the base metal being welded. Common types of filler materials include:

Welding Wires: Welding wires are used in processes like Gas Metal Arc Welding (GMAW/MIG) and Flux-Cored Arc Welding (FCAW). They are fed through the welding gun and melt to form the weld joint. Welding wires come in various materials, including mild steel, stainless steel, aluminum, and others, to match the base metal being welded.

Welding Rods: Welding rods, similar to stick electrodes, are used in SMAW. They are made of the same material as the base metal or a compatible alloy and are melted to provide additional filler material for the weld joint.

# Personal Protective Equipment (PPE)

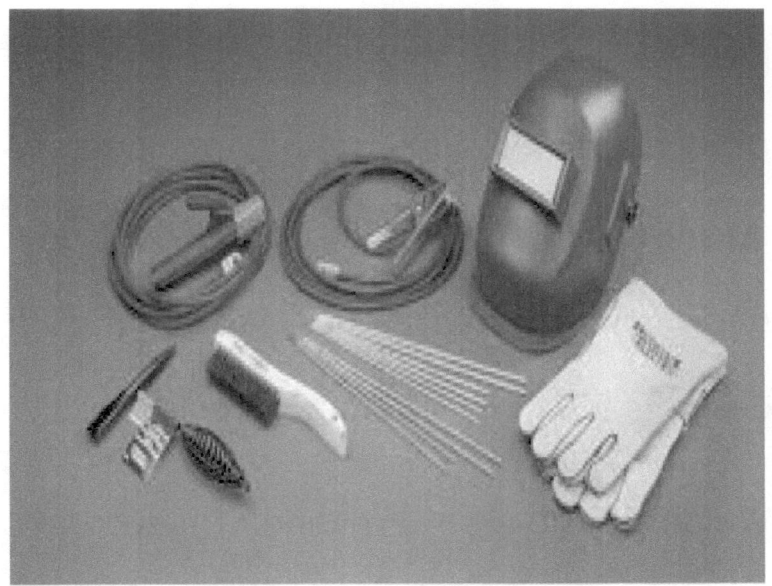

Welding poses various hazards, including intense heat, sparks, UV radiation, and fumes. Personal Protective Equipment (PPE) is essential to protect the welder from these hazards. Common PPE items used in welding include:

Welding Helmet: A welding helmet is a protective headgear that shields the face, eyes, and neck from sparks, UV radiation, and infrared light emitted during welding. It typically features a dark, tinted lens that automatically darkens when the welding arc is struck.

Welding Gloves: Welding gloves are heat-resistant gloves designed to protect the hands from sparks, heat, and molten metal splatter. They provide thermal insulation and dexterity to handle welding equipment safely.

Welding Jacket or Apron: A welding jacket or apron is a protective garment made of flame-resistant material. It covers the upper body and arms, shielding the welder from heat, sparks, and UV radiation.

Welding Respirator: A welding respirator, equipped with filters specifically designed for welding fumes, helps protect the welder's lungs from harmful gases and particulates generated during the welding process.

Welding Boots: Welding boots are sturdy, heat-resistant footwear that protects the welder's feet from sparks, molten metal, and heavy objects.

## Other Essential Tools

In addition to electrodes, filler materials, and personal protective equipment, there are several other essential tools used in welding. These tools help in preparing the workpiece, controlling the welding process, and ensuring the quality of the weld. Some of the common tools used in welding are:

Welding Clamps: Welding clamps are used to hold the workpieces securely in place during welding. They help in maintaining alignment and preventing movement, ensuring accurate and consistent welds.

Chipping Hammer: A chipping hammer is a handheld tool with a chisel-like end. It is used to remove slag, spatter, and other unwanted materials from the weld after it has cooled down.

Wire Brush: A wire brush with stiff bristles is used for cleaning and preparing the base metal before welding. It helps remove rust, paint, scale, and other contaminants that could affect the quality of the weld.

Welding Pliers: Welding pliers are versatile tools used for various tasks, such as cutting and removing welding wire, cleaning the nozzle or contact tip of a welding gun, and adjusting the position of workpieces.

Welding Hammer: A welding hammer, also known as a slag hammer or slag chipping hammer, is used to break off the slag that forms on top of the weld during certain welding processes. It has a pointed end for chipping slag and a flat end for striking.

Welding Gauge: Welding gauges are precision tools used to measure and inspect the quality and dimensions of welds. They help ensure that the weld meets the required specifications and standards.

Welding Positioners: Welding positioners are devices used to hold and rotate the workpiece during welding. They allow for easier access to different sides of the workpiece, improving welding efficiency and weld quality.

Welding Curtains and Screens: Welding curtains and screens are used to create a barrier and protect surrounding areas from sparks, UV radiation, and welding fumes. They help contain the welding operation and maintain a safe work environment.

These tools, along with the welding equipment and consumables, play a crucial role in the welding process, ensuring proper preparation, control, and inspection of the weld. Welders rely on these tools to achieve accurate and high-quality welds while maintaining safety standards.

# CHAPTER FIVE

## Welding Techniques and Joints

Some commonly used welding techniques and joints:

Butt Joint:

A butt joint is formed when two pieces of metal are joined along their edges in a straight line, creating a square or rectangular configuration. Butt joints are commonly used in welding and require proper alignment of the workpieces. They can be welded from one or both sides, depending on the joint design and welding process. Butt joints are widely used in various industries and applications, such as construction, fabrication, and automotive.

Lap Joint:

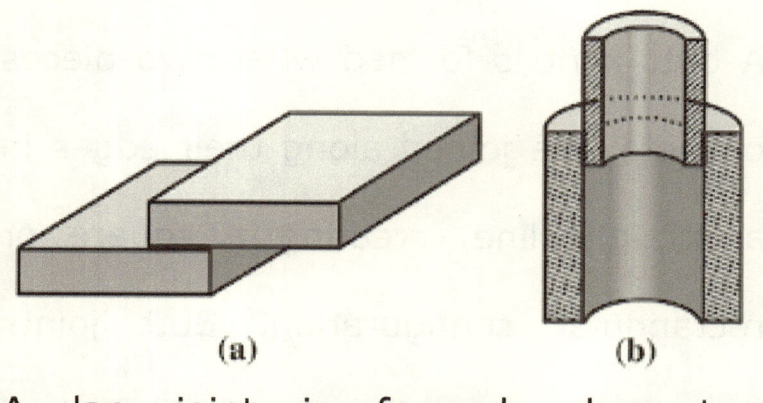

(a)                     (b)

A lap joint is formed when two

overlapping pieces of metal are joined together. One piece of metal overlaps the other, creating a joint with double thickness. Lap joints are commonly used when the primary concern is the strength of the joint rather than aesthetics. They provide good strength and are often used in applications where the joint needs to withstand tension or bending forces.

T-Joint:

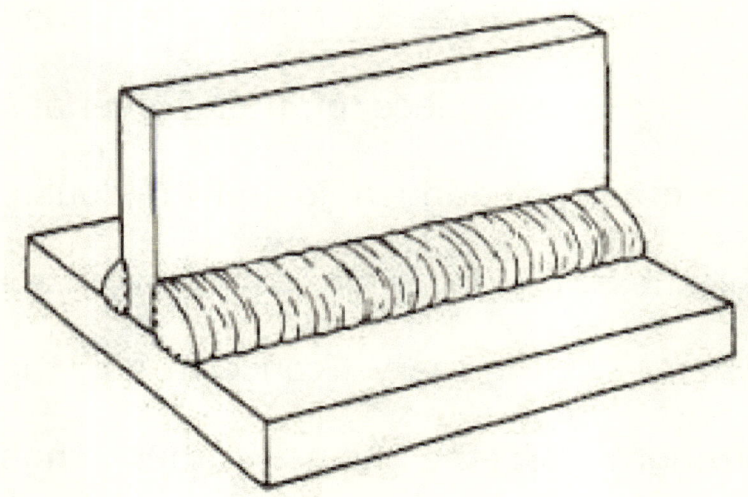

A T-joint is formed when two pieces of metal intersect perpendicularly, creating a T-shape. One piece of metal is placed perpendicularly against the surface of the other piece. T-joints are commonly used in structural applications and provide good strength in the direction of the intersecting member.

Corner Joint:

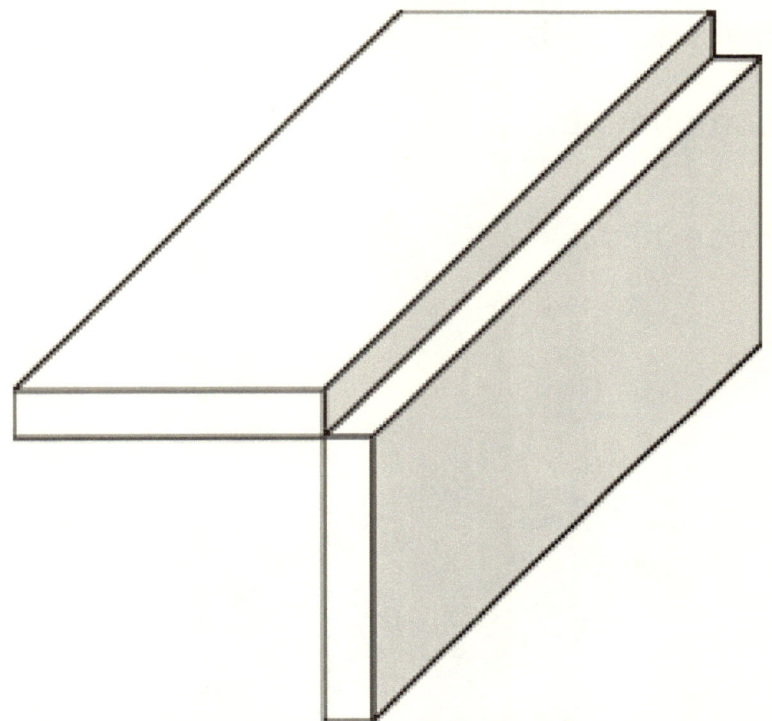

A corner joint is formed when two pieces of metal meet at a 90-degree angle, forming a corner. The weld is applied to the outside of the joint, joining the two pieces together. Corner joints are commonly used in various applications, such as the fabrication of frames, boxes, and structural components.

Edge Joint:

An edge joint is formed when two pieces of metal are joined along their edges in a straight line without overlapping. The edges are brought into close proximity, and the weld is applied to fuse them together. Edge joints are commonly used when joining thin sheets or plates, such as in sheet metal fabrication or panel assembly.

Fillet Joint:

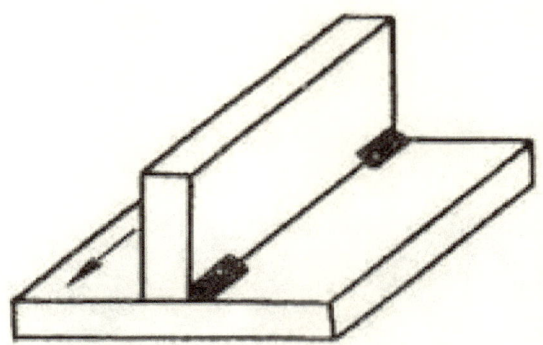

A fillet joint is a triangular weld joint formed between two metal pieces that are joined at an angle, typically 90 degrees or less. The weld is applied along the intersection of the two pieces, creating a curved or concave fillet shape. Fillet joints are commonly used in applications where strength in multiple directions is required, such as in structural welding or the fabrication of frames.

Groove Joint:

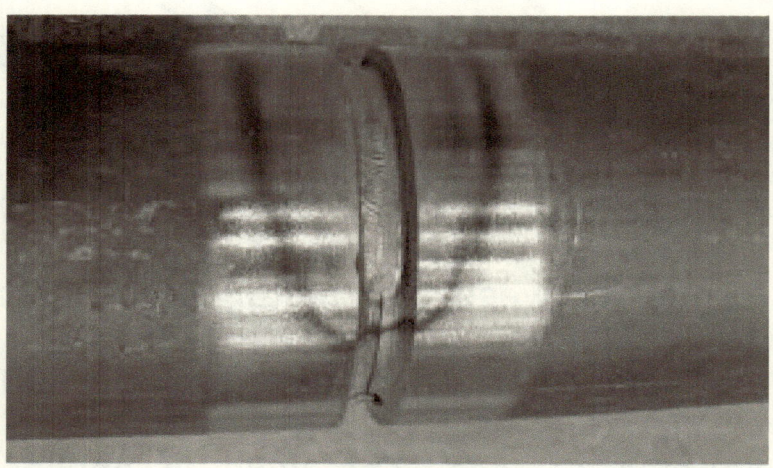

A groove joint is formed by preparing a

groove or channel in the edges of two metal pieces to be joined. The weld is applied within the groove, creating a strong and secure joint. Groove joints are commonly used for thicker materials or when high strength is required. They can have various shapes, such as V-groove, U-groove, or beveled groove, depending on the welding process and joint design.

Plug Welding:

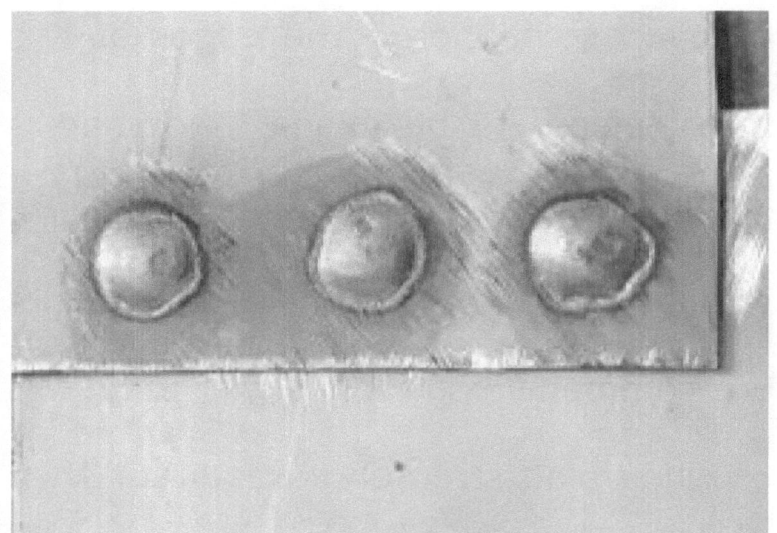

Plug welding is a technique where a hole is made in one metal piece, and the second piece is welded into the hole. The weld is applied around the edges of the hole, creating a plug-like joint. Plug welding is commonly used for joining overlapping sheets or plates, providing strength and stability.

Spot Welding:

Spot welding is a resistance welding process used to join two or more metal pieces together. It involves applying pressure and an electric current to create localized heat at the contact points, resulting in the fusion of the metal surfaces. Spot welding is commonly used in the automotive industry for joining sheet metal components.

Seam Welding:

Seam welding is a continuous welding process used to create a leak-tight joint along the length of two overlapping metal sheets. It involves feeding the sheets between two rotating electrodes, applying pressure and current to create a continuous weld along the seam.

Seam welding is commonly used in the production of tubes, pipes, and tanks.

## Common welding technique

Gas Welding: Gas welding is a welding process that uses a flame produced by burning a fuel gas, such as acetylene, in combination with oxygen. The flame creates the heat necessary for melting and joining the metal. Gas welding is versatile and can be used for various metals and thicknesses, but it is less common in industrial applications compared to other welding processes.

Flux-Cored Arc Welding (FCAW): Flux-Cored Arc Welding (FCAW) is a welding process similar to Gas Metal Arc Welding (GMAW/MIG), but instead of using a solid wire electrode, it uses a tubular wire filled with flux. The flux provides a shielding gas to protect the weld from atmospheric contamination, and it also releases additional elements to improve the weld quality. FCAW is commonly used in construction, shipbuilding, and other heavy fabrication applications.

Stud Welding: Stud welding is a process used to join a metal stud or fastener to a base metal. It involves heating the stud and the workpiece simultaneously, then applying pressure to fuse them together. Stud welding is commonly used in construction, especially for attaching studs or fasteners to steel structures.

Submerged Arc Welding (SAW): Submerged Arc Welding (SAW) is an automatic welding process that utilizes a granular flux and a continuous wire electrode. The arc is submerged beneath a layer of flux, which provides protection against atmospheric contamination and helps in the formation of a high-quality weld. SAW is known for its high deposition rates and is often used for welding thick sections in industries such as shipbuilding, pipeline construction, and heavy fabrication.

Friction Stir Welding (FSW): Friction Stir Welding (FSW) is a solid-state welding process that uses a non-consumable tool to join metals together. The tool rotates and traverses along the joint line, generating heat through friction, which softens the metal without fully melting it. FSW is commonly used for joining aluminum and other non-ferrous metals, particularly in aerospace and automotive industries.

# CHAPTER SIX

## Welding Safety and Precautions

Welding Safety and Precautions are crucial to ensure the well-being of the welder and maintain a safe working environment. Some important aspects of welding safety are:

Hazard Identification and Risk Assessment: Before starting any welding operation, it is essential to identify potential hazards and assess the associated risks. Hazards may include exposure to intense heat, UV radiation, sparks, welding fumes, electrical shock, and fire. Conducting a thorough risk assessment helps identify the necessary safety measures and precautions required for specific welding tasks.

Ventilation and Fume Extraction: Welding generates hazardous fumes, gases, and vapors, which can be harmful if inhaled. Adequate ventilation is crucial to control and remove welding fumes from the work area. This can be achieved through natural ventilation (open doors/windows) or mechanical ventilation systems (such as local exhaust ventilation). Fume extraction systems, such as fume hoods or flexible extraction arms, can be used to capture and remove welding fumes at the source.

Fire Safety and Prevention: Welding involves the use of heat, sparks, and flammable materials, making fire safety a critical consideration. Take the following precautions to prevent fires:

- Ensure a fire extinguisher is readily available in the welding area and that personnel are trained in its use.

- Clear the work area of flammable materials, such as solvents, oils, and combustible gases.

- Use fire-resistant welding blankets or curtains to protect nearby flammable objects.

- Keep a watchful eye for any smoldering sparks or flames during and after welding.

- Personal Protective Equipment (PPE): Welders should always wear appropriate Personal Protective Equipment (PPE) to protect themselves from hazards. This may include:

- Welding helmet with a darkened lens to protect the face and eyes from UV radiation.
- Flame-resistant clothing, such as a welding jacket or apron, to protect against sparks and molten metal.
- Welding gloves to shield the hands from heat, sparks, and burns.
- Safety glasses or goggles to provide additional eye protection.
- Respiratory protection, such as a welding respirator, to protect against welding fumes and gases.

Electrical Safety: Welding involves working with electricity, so electrical safety precautions are vital. Ensure the following:

- Only use welding equipment and power sources that are properly grounded and in good working condition.

- Inspect welding cables, connectors, and insulation regularly for damage and replace as necessary.

- Avoid using extension cords unless they are specifically designed for welding applications.

- Never touch the electrode or workpiece with bare hands while the welding machine is operating.

# Welding Defects and Quality Control

Welding Defects and Quality Control are important aspects of ensuring the integrity and reliability of welded structures. Some common welding defects include:

Porosity: This defect appears as small cavities or bubbles in the weld metal, caused by the entrapment of gas during the welding process.

Lack of Fusion: It occurs when there is an incomplete fusion between the base metal and the weld metal or between different weld passes.

Incomplete Penetration: This defect occurs when the weld does not fully penetrate the joint, leading to a weak connection.

Cracks: Cracks can appear in the weld metal or heat-affected zone (HAZ) due to excessive heat, cooling rates, or stresses during and after welding.

Undercut: Undercut is a groove or depression formed at the base of the weld due to excessive melting of the base metal.

Spatter: Spatter refers to the small molten metal droplets that are expelled during welding and can lead to surface contamination.

## Causes and Prevention of Welding Defects

Welding defects can be caused by various factors, including improper welding parameters, poor joint preparation, inadequate shielding gas, incorrect electrode or filler material selection, insufficient cleanliness, and operator error. To prevent welding defects, consider the following measures:

- Follow proper welding procedures and parameters as recommended by welding codes and standards.
- Ensure proper joint preparation, including cleaning, beveling, and fit-up, to promote good fusion and penetration.
- Use appropriate welding techniques and processes suitable for the base metal and joint configuration.
- Control the welding environment, including proper shielding gas coverage, to prevent contamination and porosity.

- Select the correct electrode or filler material for the specific application, considering compatibility and mechanical properties.

- Provide adequate preheating or post-weld heat treatment when required to minimize residual stresses and prevent cracking.

Non-Destructive Testing (NDT) Methods: Non-Destructive Testing (NDT) methods are used to evaluate the quality and integrity of welds without causing damage to the welded structure. Some common NDT methods for weld inspection include:

Visual Inspection: Visual inspection is the most basic form of inspection, involving a thorough visual examination of the weld and surrounding areas.

Radiographic Testing (RT): RT uses X-rays or gamma rays to create an image of the weld, revealing internal defects.

Ultrasonic Testing (UT): UT uses high-frequency sound waves to detect internal defects and measure weld thickness.

Magnetic Particle Testing (MT): MT is used to detect surface and near-surface defects by applying magnetic particles to the weld area.

Dye Penetrant Testing (PT): PT involves applying a dye to the surface of the weld, which penetrates any surface defects and is then visually inspected.

## Quality Assurance and Inspection

Quality assurance measures in welding involve various activities to ensure that welding operations meet specified requirements. This includes:

- Developing and implementing welding procedures that comply with industry codes and standards.

- Conducting welder qualification tests to assess the competency and skill of welders.

- Establishing quality control plans to outline inspection and testing procedures throughout the welding process.

- Performing periodic inspections and tests to verify weld quality and conformance to requirements.

- Documenting and maintaining records of weld inspections, test results, and quality control procedures.

# CHAPTER SEVEN

## Welding Applications

Welding finds extensive applications across various industries, including the automotive industry, construction and infrastructure, and aerospace and aviation.

Automotive Industry: Welding plays a critical role in the manufacturing and assembly of vehicles. In the automotive industry, welding is used for:

Body and Frame Construction: Welding is used to join various components of a vehicle's body and frame, such as panels, frames, and structural members.

Exhaust Systems: Welding is employed to fabricate and join exhaust pipes, mufflers, and other components of the exhaust system.

Chassis and Suspension: Welding is used to assemble and weld chassis and suspension components, ensuring structural integrity and durability.

Engine Components: Welding is utilized in the manufacturing and repair of engine components, such as cylinder heads, blocks, and manifolds.

Construction and Infrastructure: Welding is widely used in the construction and infrastructure sectors for joining structural elements and fabricating various components. Some key applications of welding in construction and infrastructure include:

Structural Steel Construction: Welding is essential in the construction of bridges, buildings, stadiums, and other structures made of structural steel.

Pipelines: Welding is employed to join pipes and fabricate pipeline systems for the transportation of fluids, gases, and other materials.

Welded Joints and Connections: Welding is used to create strong and reliable joints and connections between different structural members.

Sheet Metal Fabrication: Welding is utilized for the fabrication and assembly of sheet metal components used in roofing, cladding, and other applications.

Aerospace and Aviation: Welding plays a critical role in the aerospace and aviation industry, where precision and high-quality welds are paramount. Welding applications in aerospace and aviation include:

Aircraft Manufacturing: Welding is used in the assembly and fabrication of aircraft structures, including fuselages, wings, tail sections, and landing gear components.

Engine Manufacturing: Welding is employed in the fabrication and repair of engine components, such as turbine blades, combustion chambers, and exhaust systems.

Spacecraft and Satellite Construction: Welding is utilized in the fabrication and assembly of spacecraft and satellite structures, ensuring strength and integrity in extreme conditions.

In these industries, welding processes like Gas Metal Arc Welding (GMAW/MIG), Gas Tungsten Arc Welding (GTAW/TIG), and Electron Beam Welding (EBW) are commonly employed due to their precision, versatility, and ability to produce high-quality welds.

Shipbuilding and Marine: Welding plays a vital role in the shipbuilding and marine industry, where the construction of large vessels and offshore structures requires strong and durable welds. Some welding applications in this industry include:

Hull Construction: Welding is used to join steel plates and sections to build the hull of ships, boats, and other marine vessels.

Piping Systems: Welding is employed for the fabrication and installation of pipelines, fuel tanks, and other fluid-carrying systems in ships and offshore platforms.

Structural Components: Welding is used to fabricate and assemble structural components such as decks, bulkheads, and superstructures.

Oil and Gas: Welding is extensively used in the oil and gas industry, both upstream and downstream operations. Key welding applications in this industry include:

Pipeline Construction: Welding is crucial for joining and fabricating pipelines that transport oil, gas, and other fluids over long distances.

Pressure Vessels and Storage Tanks: Welding is employed to fabricate pressure vessels, storage tanks, and other equipment used in oil refineries, petrochemical plants, and storage facilities.

Offshore Structures: Welding is used in the construction and maintenance of offshore platforms, subsea pipelines, and drilling equipment.

Manufacturing and Fabrication: Welding is a fundamental process in manufacturing and fabrication industries, where it is used for various purposes, including:

Metal Fabrication: Welding is used to join and fabricate metal components, such as frames, enclosures, brackets, and assemblies.

Structural Steel Fabrication: Welding is employed in the fabrication of structural steel components used in buildings, bridges, and industrial structures.

Equipment and Machinery Manufacturing: Welding is utilized in the assembly and fabrication of equipment, machinery, and heavy industrial components.

Artistic and Sculptural Welding: Welding is also embraced as a creative medium in artistic and sculptural applications. Skilled welders use their expertise to create intricate and visually appealing metal sculptures, artwork, and decorative pieces.

In these industries, various welding processes and techniques are used based on the specific requirements and materials involved. Welding methods such as Shielded Metal Arc Welding

(SMAW), Gas Metal Arc Welding (GMAW/MIG), and Plasma Arc Welding (PAW) are commonly employed, depending on the application and desired results.

Each industry has its own unique welding challenges, material specifications, and safety requirements. Therefore, it is important to adhere to industry standards, codes, and regulations when performing welding in these specialized fields.

# CHAPTER EIGHT

## Future Trends in Welding Technology

Future trends in welding technology are focused on automation, robotics, and the integration of artificial intelligence (AI). Some developments and advancements in these areas are:

Advances in Welding Automation: Welding automation involves the use of advanced machinery, equipment, and systems to perform welding tasks with minimal human intervention. Future trends in welding automation include:

Robotic Welding Systems: The use of robotic systems equipped with welding tools and sensors for precise and efficient welding operations.

Automated Welding Cells: Integrated systems that combine robots, welding power sources, sensors, and programming for seamless and synchronized welding processes.

Computer-Aided Welding (CAW): Software systems that assist in the planning, simulation, and optimization of welding operations for improved efficiency and quality.

Robotics and Artificial Intelligence in Welding: The integration of robotics and AI technologies is revolutionizing welding processes, enabling increased productivity, precision, and adaptability. Key trends in this area include:

Welding Robots with AI: Robots equipped with AI algorithms can analyze real-time data, adjust welding parameters, and adapt to varying conditions for optimal weld quality.

Machine Learning and Data Analytics: AI-driven algorithms can analyze vast amounts of welding data to identify patterns, optimize welding parameters, predict defects, and improve overall process efficiency.

Collaborative Robotics: Collaborative robots, or cobots, designed to work alongside human welders, enhancing productivity, ergonomics, and safety in welding operations.

Additive Manufacturing (3D Printing): The integration of additive manufacturing techniques with welding processes, allowing for the production of complex metal components with customized designs and reduced material waste.

Laser Welding Innovations: Advancements in laser welding technologies, such as high-power lasers and beam delivery systems, enabling faster welding speeds, deeper penetration, and improved control.

Lightweight and Dissimilar Material Joining: Developments in welding techniques for joining dissimilar materials, such as aluminum and carbon fiber composites, to meet the demand for lightweight and high-strength structures.

Augmented Reality (AR) and Virtual Reality (VR): The use of AR and VR technologies to provide welders with real-time guidance, training simulations, and visualizations for improved accuracy and efficiency.

Complex Geometries: Welding-based additive manufacturing allows for the creation of intricate and complex geometries that are challenging to achieve with traditional manufacturing methods.

Material Efficiency: By adding material layer by layer, additive manufacturing reduces material waste and allows for more efficient use of resources.

Customization and Prototyping: Welding-based 3D printing enables the rapid prototyping and customization of parts, catering to specific design requirements.

Repair and Maintenance: Additive manufacturing in welding facilitates the repair and maintenance of components by adding material to damaged or worn-out areas, extending the lifespan of equipment.

High-Energy Welding Techniques: High-energy welding techniques involve the use of advanced energy sources to achieve higher welding speeds, deeper penetration, and enhanced control. Some notable high-energy welding techniques include:

Electron Beam Welding (EBW): Electron beam welding utilizes a high-velocity beam of electrons to generate heat and join materials. It enables precise and deep penetration welding, particularly for thick and complex materials.

Laser Beam Welding (LBW): Laser beam welding uses a focused laser beam to create a concentrated heat source for welding. It offers high welding speeds, precise control, and suitability for various materials and applications.

Hybrid Welding Processes: Hybrid welding combines multiple energy sources, such as lasers, arcs, and electron beams, to take advantage of their individual strengths and achieve superior welding results.

Sustainable and Environmentally Friendly Welding Methods: As environmental concerns continue to grow, the welding industry is focusing on developing sustainable and environmentally friendly welding methods. Some trends in this area include:

Low-Emission Processes: The adoption of welding processes that produce minimal emissions and pollutants, such as gas-shielded metal arc welding (GMAW) with low-fume consumables.

Energy Efficiency: Developing welding equipment and technologies that optimize energy consumption, reduce energy waste, and minimize the carbon footprint of welding operations.

Recycling and Waste Reduction: Promoting the use of recycled materials, implementing efficient material handling and storage practices, and minimizing material waste in welding processes.

## Conclusion

Skilled welders play a crucial role in the welding industry and its various applications. Their expertise, knowledge, and craftsmanship ensure the successful execution of welding projects, the production of high-quality welds, and the adherence to safety standards. Skilled welders possess a combination of technical skills, practical experience, and a deep understanding of welding processes and techniques.

The welding industry presents a range of opportunities for skilled welders. As technology advances and industries continue to rely on welding for various applications, the demand for skilled welders is expected to remain strong. By staying updated with industry trends, embracing new technologies, and continuously improving their skills, welders can position themselves for a rewarding and successful future in the welding industry.

# THE END